Magnet

Written by Darcy Patti[son]
Illustrated by Peter Willis

Moments in Science

How William Gilbert Discovered That Earth Is a Great Magnet

S N

DIAMONDS

MAGNET: How William Gilbert Discovered That Earth Is a Great Magnet
by Darcy Pattison
illustrated by Peter Willis

Text © 2024 Darcy Pattison
Illustrations © 2024 Mims House

Mims House
1309 Broadway
Little Rock, AR

MimsHouseBooks.com

For Ash,
the curious one.

Publisher's Cataloging-in-Publication data

Names: Pattison, Darcy, author. | Willis, Peter, illustrator.
Title: Magnet : how William Gilbert discovered that Earth is a great magnet / written by Darcy Pattison and illustrated by Peter Willis.
Series: Moments in Science.
Description: Little Rock, AR: Mims House, 2024. | Summary: William Gilbert wondered why the compass needle pointed north. He finally concluded that Earth itself is "a great magnet."
Identifiers: LCCN: 2024901665 | ISBN: 9781629442457 (hardcover) | 9781629442464 (paperback) | 9781629442471 (ebook) | 9781629442488 (audio)
Subjects: LCSH Gilbert, William, 1544-1603--Juvenile literature. | Physicists--England--Biography--Juvenile literature. | Geomagnetism--History--Juvenile literature. | BISAC JUVENILE NONFICTION / Science & Nature / Physics | JUVENILE NONFICTION / Science & Nature / Earth Sciences / Rocks & Minerals | JUVENILE NONFICTION / Biography & Autobiography / Science & Technology | JUVENILE NONFICTION / People & Places / Europe | JUVENILE NONFICTION / Science & Nature / History of Science
Classification: LCC QC515.G546 .P38 2024 | DDC 530/.092--dc23

William, Francis, and Thomas stared at the compass. The needle pointed north. Using the compass, Francis and Thomas could navigate the ocean, finding the way from harbor to harbor.
But why did the needle point north?

It was a mystery.

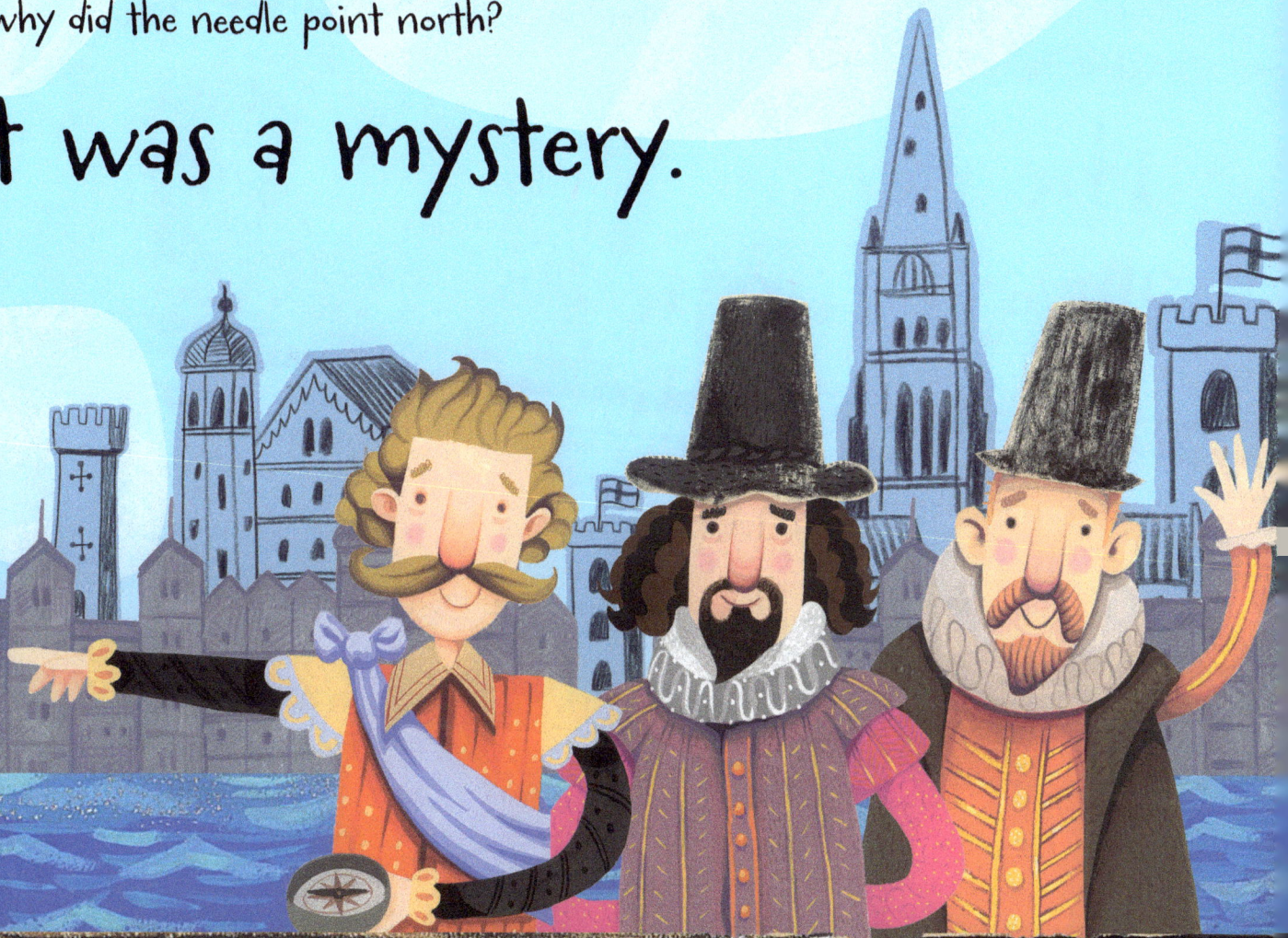

Thomas Cavendish, explorer who sailed around the world.

Sir Francis Drake, vice admiral who sailed around the world.

William Gilbert, physicist and physician to Queen Elizabeth I.

To solve the mystery, William needed to study how a compass is made. First, a long needle is magnetized, or made into a magnet, by stroking the needle with a strong natural magnet, always moving in the same direction. Then the needle is balanced on a tiny post so that it can spin around easily.

The needle always turns toward north.

William observed something else.

The compass needle dipped downward toward Earth.
Another mystery!

Mystery #1:
Why does the compass needle always point north?

Mystery #2:
Why does the compass needle dip toward Earth?

Time to study magnets.

A Magnet is a Special Kind of iron

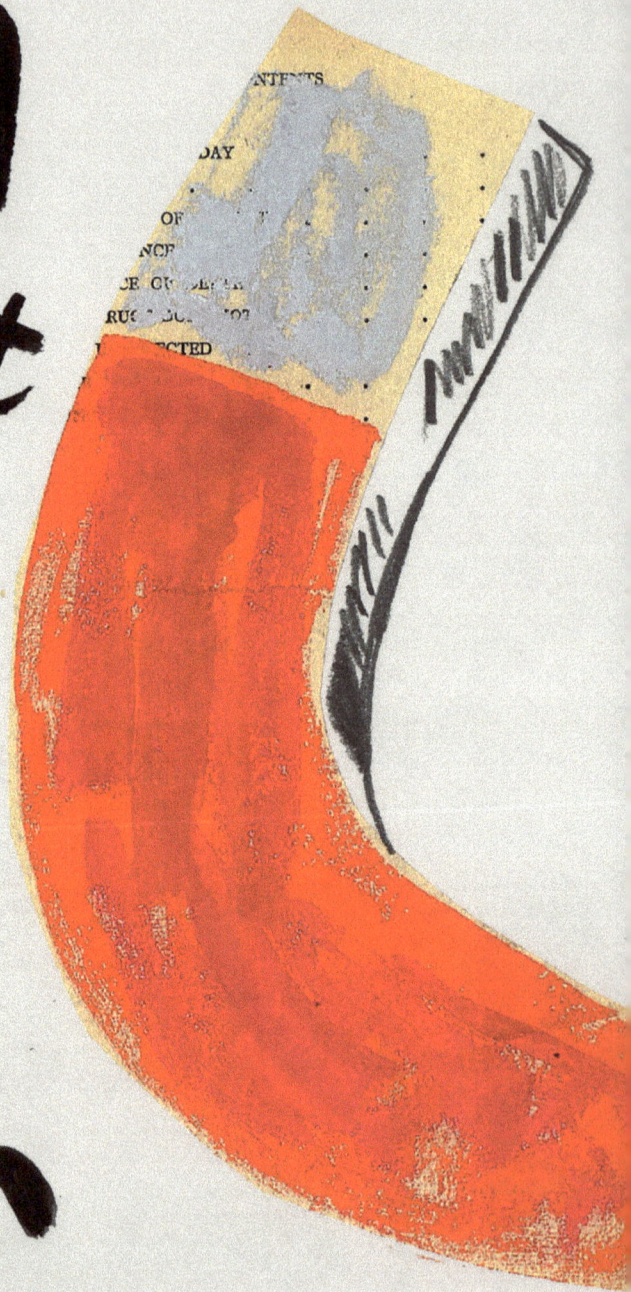

that attracts, or pulls, other things made of iron.

For eighteen years, William collected lodestones, or natural magnets, from around the world.

William wrote about how his lodestones were all different: heavy or light, strong or weak, smooth or

rough, porous (lots of holes) or solid, hard as iron or soft as clay, rocks, or clods of earth.

William heard many ideas about magnets, and his experiments showed that some were false.

Myth: Some people thought diamonds could magnetize a needle. William tried this with 75 different diamonds, and it always failed.

MYTH

Others thought that garlic weakens a magnet.

Not true.

MYTH

Some thought magnets were weaker at night.

No way.

William's experiments discovered or confirmed many other ideas about magnets.

Fact!
Magnets only attract iron. Objects without iron are not attracted by magnets.

Fact!
Magnets have a north pole and a south pole, places where the magnetic pull is the strongest. If a magnet is broken or cut in two, each new piece has its own north and south poles.

S N

After many experiments, the two mysteries were still not solved. William needed a model of Earth, a magnetic ball.

He chose a lodestone and shaped it into a ball, making a terrella, or little Earth.

William floated the terrella in a wooden bowl.

It always turned to face a certain direction. This helped him locate and mark the *terrella's* north and south poles. He also marked the magnetic equator around the middle, an equal distance from each pole.

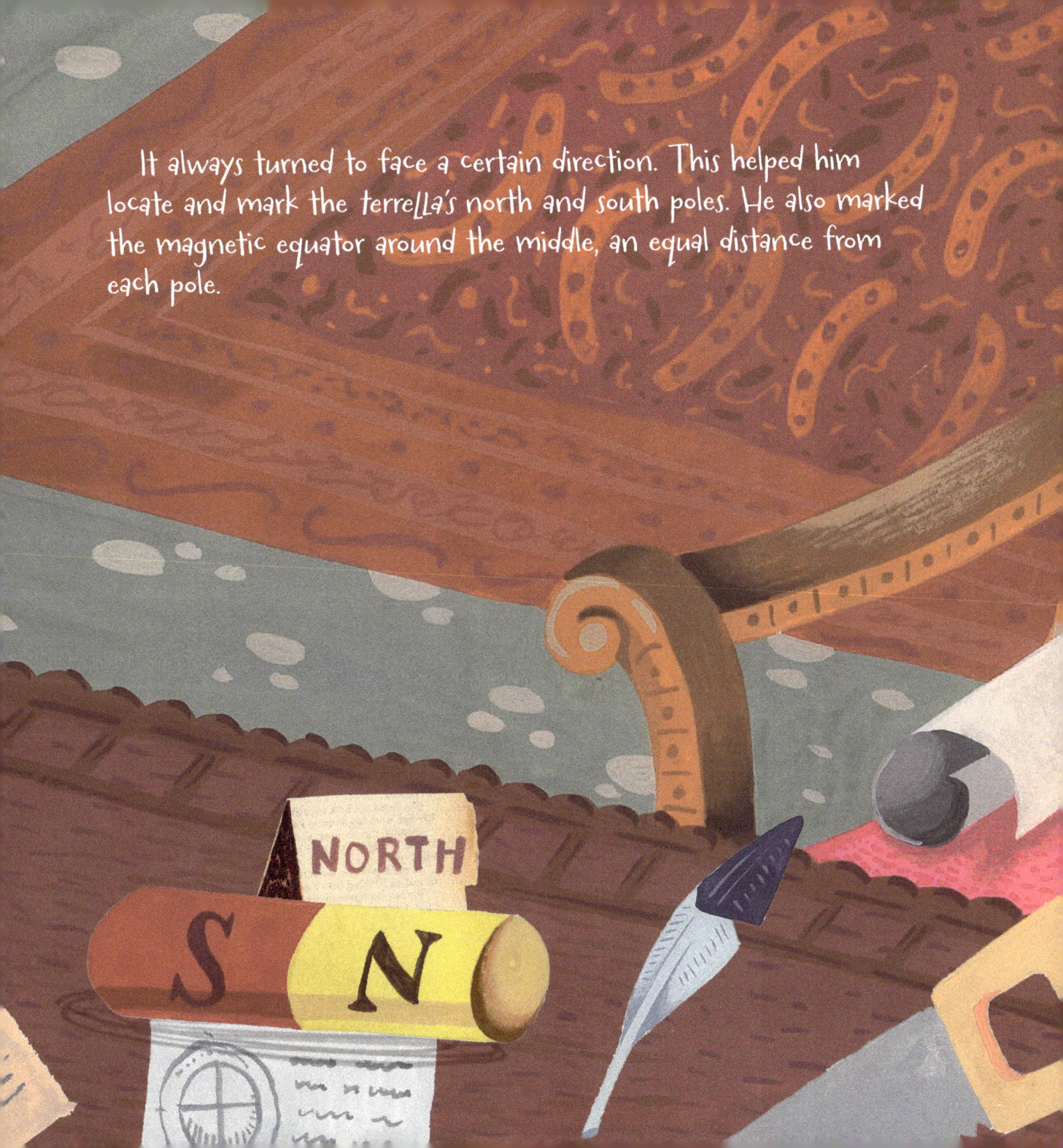

NORTH

S   N

William put a wisp of iron wire on the terrella's north pole. The wire stood out straight. It did the same thing on the south pole. But when he put the wire on the equator, it lay flat.

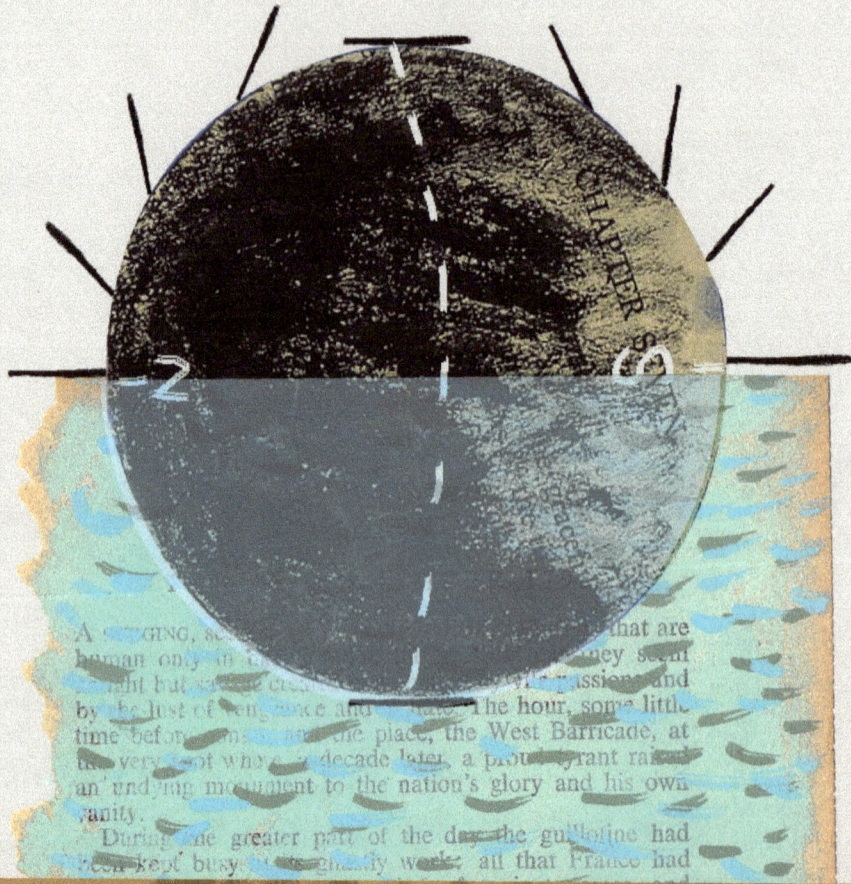

As he moved a wire from the pole to the equator, the wire started leaning. William observed that the wire was dipping
down toward
the terrella
itself.

It reminded him of how a compass needle dips down toward Earth.

The Earth-magnet was attracting the compass's magnetic needle, making it point north and downward toward Earth.

William realized something important:

Earth is a giant magnet.

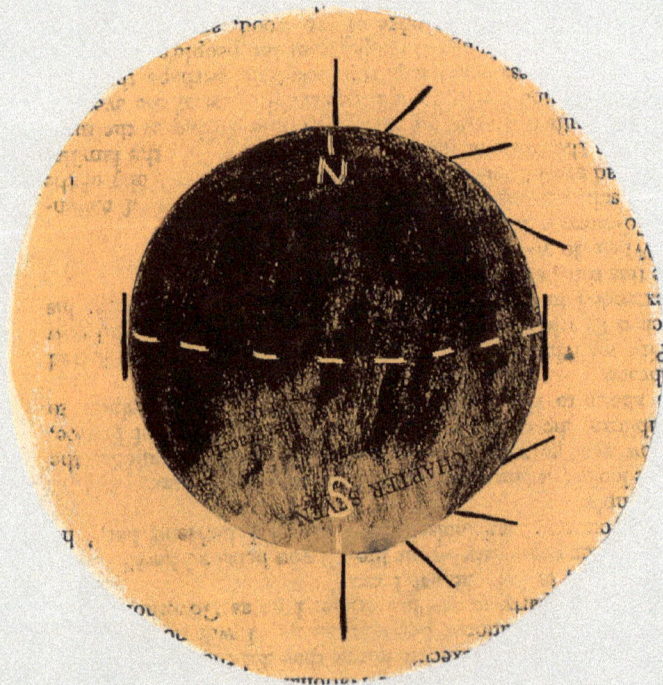

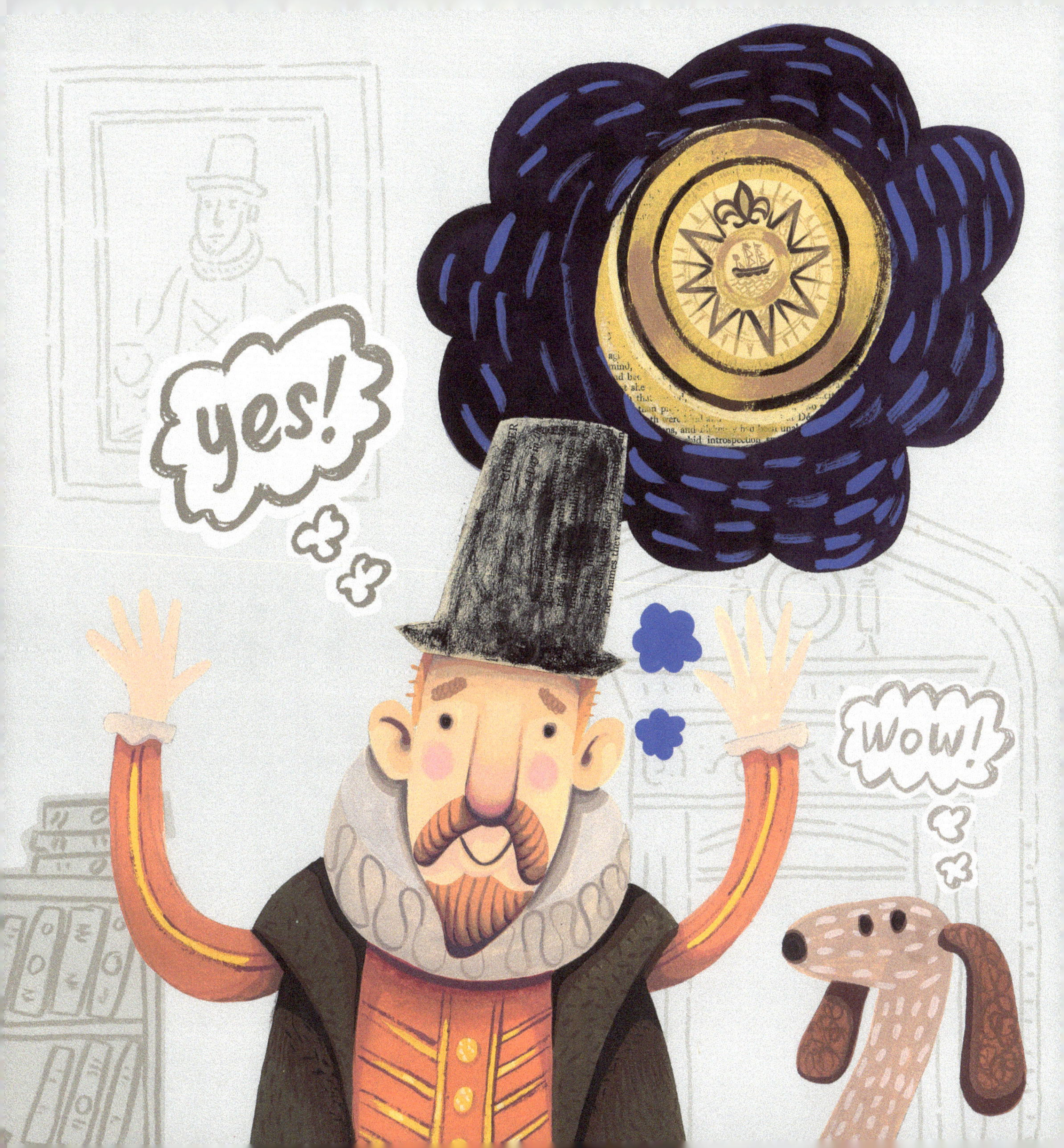
yes!
WOW!

William had solved the mystery.

After many experiments, William wrote a book about magnets called On the Magnet and Magnetic Bodies, and on the Great Magnet the Earth.

William Gilbert lived at the beginning of the Age of Science, and his magnet experiments showed others how a scientist works.

He didn't just think about things. Instead, William made observation, used models, and did experiments. Because he experimented and wrote about magnets, he is known as the Father of Magnetism.

WILLIAM GILBERT, 1544–1603 (Also known as William Gilberd or William Gylbert)

Born to a wealthy family in England, William Gilbert was the oldest of four children. But when his mother died, his father remarried, and William gained seven half-siblings. He became a physician and was elected the president of the Royal College of Physicians of London in 1600. He then became the Royal Physician to England's Queen Elizabeth I until she died in 1603. Gilbert also died in 1603, probably from the bubonic plague.

Very little is known about Gilbert. He left his papers and laboratory equipment to the Royal College of Physicians, but they burned in the Great Fire of London in 1666. Most information about him comes from his book, *De Magnete* (Latin for "On the Magnet"), or from a short biography of him in Thomas Fuller's *Worthies of England*: "His stature was Tall, Complexion Cheerful, an Happiness not ordinary in so hard a Student and Retired a Person."

Gilbert lived during the beginning of the Age of Science. Scientists of the day were thinkers, not observers. But Gilbert wanted to observe the world, not just think about it. He did experiments, made observations, and used models to understand medicine, chemistry, astronomy, magnetism, and electricity.

THOMAS CAVENDISH, 1560–1592

One of William Gilbert's friends was Thomas Cavendish, a famous explorer who had sailed around the world. He sailed west from England in July 1586, traveled below South American, crossed the Pacific Ocean, rounded Africa at the Cape of Good Hope, and returned to England in September 1588. His trip gave him many opportunities to observe a compass at different latitudes and helped inspire Gilbert to study magnets. Gilbert's book De Magnete (p. 181 of the English translation) says Cavendish and Sir Francis Drake "confirmed" Gilbert's understanding of compass needle movements.

SIR FRANCES DRAKE, c. 1540–1596

Sir Francis Drake, an English explorer and vice admiral, sailed around the world from 1577 to 1580. In England, he was known as a hero. But before and during the war between England and Spain, he pirated many Spanish vessels. The Spaniards called him The Dragon.

LODESTONES AROUND THE WORLD

William Gilbert collected lodestones from around the world, which varied in shape, size, color, hardness, density, weight, and strength. Here's how he described some of them in his book *De Magnete* (pp. 18–19 of the English translation).

EUROPE

German lodestones were perforated like a honeycomb and lightweight, but very powerful.

Central Greece mined red and black lodestones.

Norwegian lodestones were black and weak.

Spanish lodestones were white and weak.

MIDDLE EAST

Lodestones from Arabia were red and shaped like tiles.

Lodestones from Asia Minor (Turkey) were black and weak.

ASIA

Lodestones from the East Indies, China, and Bengal (India) were dark blood-red, often of great size and weight and were strong enough to lift up to a pound of iron.

Siberia had the strongest lodestones.

AFRICA

Ethiopian lodestones were amber-colored and strong.

SOURCES

Gilbert, William. On the Loadstone and Magnetic Bodies, and on the Great Magnet the Earth. Translated by P. Fleury Mottelay. London: Bernard Quaritch. 1893. https://archive.org/details/onloadstonemagne00gilbuoft

Landgon-Brown, W. William Gilbert: His Place in the Medical World*. Nature 154, 136–139 (1944). https://doi.org/10.1038/154136a0

USING A MODEL TO UNDERSTAND MAGNETISM

Terrella, c. 1661 © The Royal Society

When Gilbert was trying to understand magnetism and how it affected a compass, he decided to create a three-dimensional model of Earth, a *terrella*. The model let him visualize, or see, his ideas about magnets and compasses.

To create the model, Gilbert chose a strong lodestone and shaped it into a sphere. Gilbert observed the behavior of wires at different places on the sphere. He compared that to the behavior of a compass needle at different places on Earth. That helped him deduce that Earth is a great magnet.

Models like the terrella are an important tool that enable scientists to understand the world around us. Models don't explain everything, however. Gilbert didn't know why Earth was a magnet or how it all worked. Later scientists discovered magnetic fields and helped build a fuller understanding of magnetism.

WHY IS EARTH A MAGNET?

Scientists still can't explain why Earth is a magnet, but they believe that Earth's molten core may be partly responsible, but they still don't know. Across our solar system, most planets are magnetic. Only Venus is not magnetic and never has been. Mars and the Moon are not currently magnetic, but they have deposits of magnetic metals.

FERROMAGNETIC ELEMENTS

William Gilbert thought only iron could be a magnet. Today, we know that other metals can also become magnets, either temporarily or permanently: iron, nickel, cobalt, and some rare metals.

Moments in Science

BURN — Michael Faraday's Candle — Darcy Pattison, Peter Willis

CLANG! — Ernst Chladni's Sound Experiments — Darcy Pattison, Peter Willis

POLLEN — Darwin's 130-Year Prediction — Written by Darcy Pattison & Illustrated by Peter Willis

ECLIPSE

EROSION

P.I.

FEVER — How Tu Youyou Adapted Traditional Chinese Medicine to Find a Cure for Malaria — by Darcy Pattison, Illustrated by Peter Willis

AQUARIUM — How Jeannette Power Invented Aquariums to Observe Marine Life — Written by Darcy Pattison & Illustrated by Peter Willis

Magnet

MOMENTS IN SCIENCE — Learn More

*FOR THE KIDS WHO NEED TO KNOW—
BECAUSE THEY GROW INTO THE BOOKS THEY READ.*
—Darcy Pattison

···ANOTHER EXTRAORDINARY ANIMAL···

Bird — WISDOM THE MIDWAY ALBATROSS — Darcy Pattison

Mammal — ABAYOMI THE BRAZILIAN PUMA — Darcy Pattison, Kitty Harvill — The True Story of an Orphaned Cub

Marine Mammal — PELORUS JACK THE NEW ZEALAND DOLPHIN — Darcy Pattison — Inspiring a Government to Protect an Individual Animal

Spider — NEFERTITI THE SPIDERNAUT — Darcy Pattison — The Jumping Spider Who Learned to Hunt in Space

Amphibian — ROSIE THE RIBETER — Darcy Pattison, Nathaniel Gold — The Celebrated Jumping Frog of Calaveras County

Reptile — DIEGO THE GALÁPAGOS GIANT TORTOISE — Darcy Pattison, Amanda Zimmerman — Saving a Species from Extinction

Extraordinary Animals — Learn More